T

Firefighter Preparation Beyond The State Skills Test

Captain Collin S. Blasingame
Lieutenant Justin C. Dickstein
Firefighter John E. Gomez

Copyright © 2013 by Powder Keg Publishing, LLC
All rights reserved.

ISBN-13: 978-0615760971
ISBN-10: 061576097X

No part of this book may be reproduced, scanned, or distributed in any printed or electronic form without permission. Please do not participate in or encourage piracy of copyrighted materials in violation of the authors' rights. Purchase only authorized editions.

Table of Contents

Introduction 5

Section 1: At the Station 9
  I. Be Proactive 11
  II. Get Ready to Clean 15
  III. Know the Tools 17
  IV. It is Your Equipment 21
  V. Learn to Cook 25
  VI. Use Your Time Wisely 29
  VII. Be on Time (Early) 33
  VIII. Bring Money 35
  IX. Look Like a Professional 37
  X. Get in Shape 39

Section 2: On Scene 41
  XI. Learn Something on Every Call 43
  XII. Take Responsibility for Your Own Safety 47
  XIII. Wear Your SCBA 51
  XIV. Smooth is Fast 55
  XV. Do Not Freelance 59
  XVI. The Right Tool for the Job 63
  XVII. Fight Fire Like You Have Been Trained 69
  XVIII. Learn Building Construction 73
  XIX. Learn to Read Smoke 77
  XX. Get Ready to Drive 79

Section 3: Attitude Is Everything 83
  XXI. Take Pride in Your New Career 85
  XXII. Learn the Rules 87
  XXIII. Respect Your Elders 91
  XXIV. Listen More Than You Talk 93
  XXV. Do Not Be Afraid to Ask Questions 97
  XXVI. Learn From Your Mistakes 99
  XXVII. Train 101

XXVIII. Actions Speak Louder Than Words 103
XXIX. Watch Out for 10/20 Syndrome 107
XXX. Leave Your Opinions at Home 109
XXXI. No Temper Tantrums 113
XXXII. Learn to Love EMS 115

Conclusion 117

Feedback 119

About the Authors 121

Introduction

This book was written to help new firefighters know and understand what will be expected of them when they are first assigned to a fire station and throughout their probationary period; as well as assist fire departments and training academies effectively prepare their personnel for all aspects of station life. This is one of the most exciting, and stressful, times in a new recruit's life. For someone who has always dreamed of, or is even just now considering, becoming a firefighter, this book paints a realistic picture of the expectations a new recruit faces. It includes all the details, from what to do around the station to some suggestions on how to be more successful at an emergency scene. Most rookies have to discover all of this through a painful process of trial and error. This book is designed to shorten that learning curve and to help the reader

Introduction

become an outstanding new firefighter from day one.

The fire academy does an excellent job of teaching the technical skills and the fundamentals of firefighting. However, one of the most overlooked areas in the fire service is life as a rookie. Be it due to time, budget, curriculum, or any number of other restraints, how to be a good rookie is a subject that is generally either quickly skimmed through or ignored altogether. This is a great disservice to both the hopeful firefighter and his or her future department. Many times expectations for the new recruit are never carefully laid out. This lack of communication can quickly lead to a perception of unacceptable performance through no fault of the newly hired rookie. And once the reputation of poor work habits has been established it becomes a self-fulfilling prophecy, as the rest of the department is quick to spot inadequacies and point out deficiencies. This can quickly spiral out of control, making the rookie extremely uncomfortable, and cause the fire department to regret their hiring decision. It could very likely cost someone a chance at one of the greatest careers in the world. This can all easily be avoided with a simple lesson in expectations, which is exactly what this book is. It should be used as a guidebook to make sure the rookie knows all of his or her duties. By following the advice laid out in this book, the new recruit can confidently show up on his or her first day at the station knowing that they have a wealth of information to become a great rookie firefighter that any station in any department would be proud to work with.

This book is also a helpful tool for someone who is considering a career in the fire service. There are a lot of misconceptions of what life at the station and in the fire department is like. It can be a truly rewarding and exceptionally enjoyable career. But, it is unlike any other profession. The new rookie must prove himself or herself before they are accepted into their new "family". This book will give the reader a clear picture about what it takes to do just that. It is not a book about glamorous rescues and running into burning buildings, it is a real look at all of the dirty work that is expected during the first year on the job. These are the duties that lay the foundation for an exhilarating, inspiring, and fulfilling career in the fire service.

Section 1: At the Station

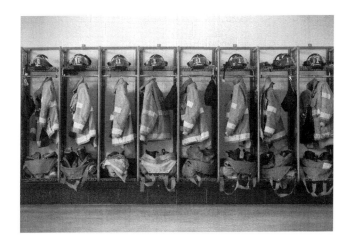

I. Be Proactive

When you arrive on shift, everything will be new to you. You will begin a new routine, the calls will be real, and you will be introduced to station life. Slow and fast, patterns will develop. Some things you will be told, other things you will see, especially if you are paying attention. When you are told something, make a good effort to remember it. If needed, write it down. If you have a question, ask. The more things you can anticipate, the better you will be at this job. When you have an idea of what is coming next, you can take the next appropriate step. Knowing and implementing this will change you from a liability to an asset. When you are new and do not know much, you are a liability. This is okay when you are starting out; it is expected. However, you should strive to become an asset, and that comes when you are proactive and thinking ahead. At

At the Station

that point you can be counted on to do what needs to be done, without someone dictating your every move.

Becoming proactive around the station should come very fast. This happens because tasks are not complicated and the ramifications of a mistake are not very high. For instance, once you know what day the yard is mowed or degrease is performed on the apparatus, take the initiative and begin the task without being told. After all, you would rather be criticized for poor judgment than being lazy. Your pace on calls will be a little slower. Every call is different, but they have their similarities, and what you learn on one call can usually be applied in some form to another call. The next time you respond to a similar emergency, you should know what tools to grab without being told. Additionally, in the morning you will probably hear about the other shifts and the calls they had. The majority what is said is intended to give someone a good ribbing. Yet, it is also supposed to share a wealth of knowledge: what the call was and how it was handled.

As a rookie, being proactive is constant. For instance, look for the dirty work. When you see the crew is about to start on a task, ask yourself, "What is the worst job here?" The answer to that question is the job you need to be doing. After dinner, you need to be the one washing the dishes. If you are loading hose, you need to be in the hose bed. If you are mowing the yard, you should grab the oldest, hardest to use mower. This should be done without anyone telling you to do so, and it should be applied to every moment of your rookie year.

II. Get Ready to Clean

Your expertise should be in cleaning. As a rookie the one job you should require the least amount of training in and are expected to do continuously is to clean. The station should be clean at all times, inside and out. This falls directly upon your shoulders. This starts the moment you arrive at work and lasts until you leave to go home. Yes, you are going to clean up after the shift before you, and yes, you are going to clean up for the shift after you. Get over it. It is like going camping; leave it in better condition than you found it. If at any time you discover yourself looking for something to do, clean. Your work ethic can and will be determined by how clean the station is.

At the Station

Nowhere is it more apparent than the restroom. If you think cleaning toilets is beneath you, perhaps you should find a new profession. Every restroom should be cleaned at least twice a day, once in the morning during station cleanup and again during the early evening. Once clean, it should be obvious. The restroom should smell of bleach because you used water with bleach to mop. The toilets need to be scrubbed and wiped down, inside and out, including the handle. They should have clean soapy water in them when you are finished to let everyone know that they are clean. Do not use the powder detergents, it collects in the bottom of the bowl and clogs the pipes. One particular rookie put a strip of clean toilet paper across the toilet seat, tucking in the edges as if it was a nice hotel. It made a good impression that I still remember to this day. The trash should be emptied, the sinks wiped down, paper towel dispensers filled, and the mirror should be free of any spots. Everyone likes a clean restroom; make sure they get what they want.

A thorough cleanup of the entire station should be done every morning. You should be shown what needs to be done your first shift. Make sure you do it and more. If at any time something becomes dirty, you need to clean it. Keep an eye out throughout the day for anything that needs touching up. It is not uncommon for someone to leave something amiss just to see how long it takes to get cleaned. Station cleanup is your job; make sure it is always clean.

III. Know the Tools

Think of this scenario. You have been on the department for several years; you have some experience under your belt, and have seen a few things. You are driving down the road in a rent-a-car, without your phone and the car is empty. One hundred yards up the road, a terrible accident happens. People are in trouble. What could you do? The unfortunate answer is not much, and the reason is because you do not have the necessary equipment. Sure, there are a few things you might be able to do, but you will not have the impact you would like simply because you are missing some key elements: your gear, crew, and equipment. If you do not know your gear, you might as well not even have it.

At the Station

As a rookie one of your first and most important jobs is learning your equipment. That includes everything you can put your hands on. And it is going to take exactly that, putting your hands on it. I have told rookies this before, face to face, and for some reason they think seeing it, or pulling out the container the tool is in, is putting their hands on it. It is not. Picking up the individual tool, ensuring that it is in good condition, and knowing how to operate the tool is putting your hands on it. Your fingerprints or "gloveprints" should be all over every tool on the apparatus. If you have a question about a tool, which you should, ask the driver of the apparatus, read the manual, call the manufacturer, or do whatever it takes to find out. You should know the specifics on the tools better than anyone else. Know what they can do and what they cannot. You might not be able to handle them as well on scene as some of the veterans, it requires some experience, but you should know the details. That being said, an emergency scene is not where you become familiar with a tool, it is where you expand your knowledge. The location of every tool on the apparatus should be known before it is needed. If your officer asks for a tool (after a couple of calls of the same kind, he or she should not have to ask) you should only need to open one compartment. Smooth is fast, and time is important in an emergency.

This principle still applies when you swing; your equipment is on the apparatus you are riding that day. When you get to a station look through the compartments and learn the tools. Throughout the day, look through the compartments again. Learning your equipment takes some time, but you owe it not only to yourself, but also to your family, your crew, and the citizens of the community to have a thorough knowledge of the gear.

IV. It is Your Equipment

As you walk in the door at the station the very first place you need to go is to your gear locker. Pull out your PPE and inspect it. Sure, it will probably be as you left it, but it has been out of your care since you went home. Perhaps someone lost a glove and borrowed yours. Maybe you accidentally hit your voice amplifier, turning it on and now it is dead. At this point you do not know, so check it. Make sure your face piece is in proper order and still able to provide a proper seal. If your department does not issue a bag to store your face piece, buy one and use it. If a glove fails, your hand might get burnt. If your face piece fails, things are going to be much worse. After you ensure that your gear is in proper working condition place it beside the apparatus you will be riding. Do not remove the person's gear you are

At the Station

relieving. While this may seem like a nice thing to do, you will find that <u>most individuals do not like their gear being touched by others</u>.

<u>Never put anything on the bumper or on a step of an apparatus, you will lose it.</u> It is a common mistake that does not have to be learned the hard way. If you put something there, consider it gone. As a rookie, I was mowing the yard and riding on any apparatus that ran. At the time, we had two different engines at the station. One caught an automatic fire alarm call, I ran in grabbed my gear off the other engine and got bunked out. In the process, unbeknownst to me, I managed to kick my face piece under the engine. As we headed down the street I tried to get fully bunked out, but I could not find my mask. Luckily, we were disregarded. When we pulled back into the station and the garage door opened, my mask was there. The problem was, it was in many more pieces than it should have been, having found its way under a tire. I got another mask within the hour, I then was tactfully told to wear it the rest of the shift. So I did. I learned something that I already knew, but did not properly execute, gear accountability.

Every shift for your entire career you need to check your air pack. As a rookie, you need to check everyone's air pack. You should check everything you can on the apparatus. This practice provides you with a good opportunity to become familiar with the equipment and lessens the burden on the driver. He or she will be taking plenty of time out of their schedule to teach you, so help them when you can. This will include changing the batteries in all the portable radios, checking that there is water in the tank, fuel in the power tools, refreshing the ice water in the cooler, checking the combustible gas detector and CO detector, putting fuel in the apparatus, plus anything else that can be done. Everything you do, or fail to do, must be reported to the driver of the apparatus. If they check it again, do not get offended, it is their responsibility and their name on the dotted line. Every time you pull into the station you need to ask yourself what was used. It is easy to forget to refill the absorbent when it was three calls ago and you have been responding to various incidents for two hours. This mishap can be avoided by asking yourself that question every time you set foot off the apparatus. Do this immediately upon returning because there is no time like the present. It does not matter that dinner is ready and on the table. Take care of your equipment first, so it can take care of you.

V. Learn to Cook

A large portion of your career will be spent at the station. There are plenty of guys and girls within the fire department that are competent at their job, that know what to do on a call, yet they still end up on a good portion of transfer lists. The reason for this is usually because they fail to get along at the station. A sure fire way to increase your value is by preparing good quality meals. Meals are a very important aspect of station life. The kitchen is where stories are swapped, trash is talked, misunderstandings settled, information learned, and countless other things transpire. Everyone eats, and if you like to eat, you should know how to cook. It is that simple. As a rookie, you need to cook or help cook all the meals.

At the Station

If you do not know how to cook, you need to learn. There are plenty of resources for this. The best single way to learn is from a good cook. Maybe it is a parent, spouse, grandparent, mother-in-law, or the station cook, whomever, ask them to teach you and help them prepare some meals. You are going to need to do this more than once. It takes time to be proficient. Make it a weekly routine at the least and take notes. If you do not know any good cooks, go to your local library, bookstore, or browse the web and get some recipes. You will need to fully read the recipe before you start and prepare the items you can immediately beforehand: chopping the onions, peeling the garlic, etc. Do not cut corners; a good meal is made in the details. Another good resource is videos, whether it is a cooking show or a home video off the internet. Of course, the best way is a combination of the above. You always need to prepare the meal at least once at home before you cook it at the station. Your family and friends will be your lab mice. A good meal will get a nod at best; a bad meal will be talked about your entire career.

One of the most important aspects of cooking is staying within budget. If you have fifty dollars to feed six firefighters, filet mignon is off the menu. At first, keep is simple. Start from scratch, it cost less and tastes better. That is to say, mashed potatoes come from whole potatoes, not from a box with water added. Although you need to stay within budget, you also need to make enough food. There should be leftovers; you never want to run out. Make more of the items that cost less. When planning the meal, try to keep other people's diets in mind. The days of fried food for every meal have gone away. You should not have to cater to every vogue diet, but you cannot fatten the veterans for early retirement either. Learning to cook will pay dividends not only within your fire department career, but also for the rest of your life.

VI. Use Your Time Wisely

Everyone has the same amount of time in the day, twenty-four hours. What you get accomplished is based on how you allocate that time. It is no secret; successful individuals manage their time well. As a rookie you have an enormous workload and it will require a large portion of your time to become proficient at your job. Using your time wisely will not only help you become better at your job, it will provide you with more opportunities to spend with your family and loved ones. One simple way to start is by creating a list of the things you need to get done every day. This list is dynamic, things will get added and things will get removed. I personally make mine the night before and cross out anything I accomplish. If something comes up that I do immediately, I write it on my list just to cross it out. It lets me

At the Station

know what I got done at the end of the day and what I still need to do.

Additionally, a way to get more done is to not waste the time you do have. This includes not watching television. Your first year you should not watch any television unless your crew demands it. You simply have too much to do, not to mention it looks bad. Likewise, stay off your phone. A call to your spouse after you wash the dinner dishes is expected and encouraged, constantly texting and updating your social media page is not. You should check your city email, but should not surf the web while doing so. After your rookie year is up you will have plenty of time to do these things, but right now you need to concentrate on learning the fundamentals of your job.

You can learn the fundamentals by incorporating good study habits into your daily routine. Let us say you need to learn your streets in your district, which you do. On any call you do not need to bunk out, look at the address on your map and think how you would get there. Now watch how the driver gets there. Is it the same? If not, why? When you leave work, drive through a couple of the streets you have not been on. Little things like this do not take much time, but provide great rewards when consistently done.

This job evolves over time; you need to evolve as well. Saying something is new, will not provide you an excuse of why you do not need know it. Not being able to fill out an electronic patient care report is the same as not being able to start a chainsaw. Regardless of your opinion, it is on the apparatus and you still are responsible for it.

VII. Be on Time (Early)

If you are not early, you are late. One of the fastest ways for you to develop a poor reputation is to be late. You need to be the first one on your shift at your station every shift unless told otherwise by your officer. If your shift starts at 0700, you should probably be there by 0600. Being this early allows for the unexpected: a flat, a dead battery, traffic, lost keys, etc. Being late once in a career is acceptable, being late once in your first year is not. The last thing anybody wants to do is wait for their relief after they have been working for twenty-four hours with only three hours of sleep. If they do, you can guarantee they will have words for you and everyone else about you. If this means you need to have three alarm clocks with two being away from your bed, get four. You might need to set everything out the

At the Station

night before to facilitate the morning routine or load your vehicle the evening before. It is that important.

At any point when you need to be at a certain place at an appointed time, be there beforehand. No one should have to wait on you during your rookie year. If for some reason you are going to be late, call. Most problems can be lessened with good communication. The positive side to always being early is that if there is a time when you need someone to hold over for you, you should not have a problem finding someone to do it.

VIII. Bring Money

One particular firefighter I worked with never had his money for meals. He always had an excuse, but it was never accepted as payment at the store. He would always pay, but it was never up front. It got to the point, and it did not take long to reach this point, that a rule was established. If you did not have your money at the beginning of the each shift, you automatically volunteered to cook and buy ice cream. Amazingly, after this rule change he always had his money. Go figure.

You have a job; you get paid to perform this job. Everyone you work with knows how much money you earn. <u>Have your money</u>. If the food kitty is ten dollars every shift, as a rookie have the correct change in <u>bills</u>. Do <u>not</u> bring <u>coins</u>, do not bring an excuse, just bring your

At the Station

money. You should pay before you sit down to eat. If you swing, ask how much the food kitty is, then pay immediately. You should also put twenty dollars in your bunker pants. At some point you will be out on a call on a hot summer day and that money will buy pure happiness, an ice cream or drink for everybody.

In your budget, allocate more than just the food kitty. There will be fundraisers, Girl Scout Cookies, band chocolate, among other costs. Every once in a while, on special occasions you will need to buy quality steaks. On other occasions you should buy donuts or ice cream. In the end, you get what you give. Sure, there will be that individual that does not pay his or her fair share, and he or she will be begrudged by all that do. Do not be that person; it costs less just to pay the money.

IX. Look Like a Professional

Everyone has heard that you should not judge a book by its cover. Perhaps, but that cover is a part of the book and the first thing you see, so you can bet it will be judged. The first way you are going to be sized up is by your appearance; you need to make it a good one. Every day as a firefighter you will come into contact with people you have never met, as a rookie this is especially true. Your uniform needs to be clean and pressed; your boots need to be polished. Your uniform should fit well. If it does not, see about getting some that do. Always have at least one change of clothes ready at all times and as a rookie you need at least two. It only takes one call, one prank, or one slip with the mop bucket to need them. Take pride in wearing your uniform, not only do you represent yourself and your family; you represent

At the Station

the fire department and every other firefighter that has put on a uniform.

    Stand up straight with your shoulders back; look at people in the eye when you speak to them and when they speak to you. Offer a firm handshake when you meet your fellow coworkers. Get a regular haircut and always be clean-shaven when you arrive at work. Brush your teeth after every meal, and keep some gum handy for those late night calls. Take a shower every shift. And yes, eventually you will get caught in the shower with lather in your hair, it happens. At that point, you rinse extremely fast, throw on your clothes and grab your socks (it takes longer to try to put wet feet into turnout boots than to put on socks then the boots, trust me), and hop on the waiting apparatus. In short, you always need to be ready to make that first impression because it is around every corner.

X. Get in Shape

Firefighting is a physical job. To perform your job well and to have a full career you need to be physically fit. There is no way around this simple truth. After any amount of time in the profession, you will come across the individual that either lacks the self-discipline or physical capability to keep in good shape. They dread a fire tone and become a liability to themselves and the people around them.

Your health is the most important responsibility you have. You need to do what you can to take care of yourself. It starts with a workout program. You need to start a strength-training program with cardio exercises. Being able to lift a house is of no use if you cannot run twenty-feet. Likewise, being able to run a six-

At the Station

minute mile does little for if you do not have the strength to control a hose line. Consistency is the key element with any workout regimen. Start your program now; it is much easier to stay in shape than it is to get in shape.

Make smart decisions regarding your diet. Choose a well-balanced meal, and take it easy on the sweets. The most important exercise you can learn might be pushing yourself away from the table. Every choice you make with regard to what you put in your body has its consequences, including tobacco and alcohol.

When it is time to sleep, do just that. This job will thwart your sleep enough without any additional help from you. There is no need to waste an hour of sleep watching a sitcom with its accompanying commercials that you get nothing out of. When you lay down in bed, turn out the lights and go to sleep. To do otherwise is rude to others in the room, and it will only negatively affect your performance the next day. It might be hard to fall asleep at first, but your body will get use to it and thank you later.

Section 2: On Scene

XI. Learn Something on Every Call

Few careers in the world are as varied as the fire service. Every time you walk into the fire station to start your shift you have no idea what the next twenty-four hours will hold. You may perform CPR on someone, fight a warehouse fire, respond to a train derailment, or a tornado may hit your town; you just never know. To be honest, there is also no guarantee that you will ever walk out of there when the shift is over. To say you must constantly be a student of your profession is an understatement. Too many firefighters think they know everything. This should never be the case. Education must continue your entire career. Learning something new every shift is good, but something can be gained on every call. This is not as difficult as it sounds. It simply comes down to being observant, preplanning, and critiquing the call in your head afterward.

Be observant on the way to the call. Learn the buildings and streets in your district. Notice the weather and wind patterns while running scenarios through your head. If you pull into a neighborhood, learn the neighborhood. Some neighborhoods have vacant streets during the day and cars on both sides of the street at night. Notice how visible the house addresses are. Would they be visible at night? Note the hydrant spacing on the streets. Are there dead-end streets? Are the cul-de-sacs large enough to turn the apparatus around in? This information will help you when it is your turn to drive. Look at the houses as you pass them. What are the general age, size and construction of the houses? Have most been added on to? Do the houses have burglar bars? Does the same floor plan repeat every third house? Are the utilities underground or overhead? Does the neighborhood have natural gas service? Can the garages be accessed from the street or alley? You will need to know this information in an emergency.

Preplanning is also a great way to learn each shift. You should mentally preplan every structure you walk into. It doesn't matter if it is a residential house you are responding to for a medical call or the local grocery store you buy dinner at. Start outside by noticing the visibility of the business name or address and the access points. If the structure is commercial, look for the

Knox box, utilities, FDC connections, placards, etc. While still outside, try to visualize the layout of the home or building and the construction type. Once you are inside, take inventory again. Did it match how you pictured it? Would you be able to navigate inside with heavy smoke conditions? Are there any hazards present which were not visible from the exterior? This is a great exercise that will pay dividends.

Another great method to learn by is critique. You should mentally evaluate every call you go on. Every incident has things that go well, and things that could have been done better. After the incident is over and the adrenaline has left, mentally replay the call in your head. Look at everything from response time to customer service. Pick up the good habits of others, and learn from mistakes. Remember, your mental critique is personal and for your learning only. Keep it to yourself, and do not judge others or point out their mistakes to them or anyone else.

The fire service is a very noble and prestigious career. However, with that comes great responsibility. Now that you are a firefighter, you are responsible to your family, your crew, your employer, and to those you serve to be the best firefighter you can be. One aspect of being the best is to constantly gain knowledge of your job. Knowledge is gained by critiques, preplanning and simply being observant. Never stop learning.

XII. Take Responsibility for Your Own Safety

Each year close to one hundred firefighters die in the line of duty and thousands more are seriously injured. No one will argue that firefighting is not a dangerous line of work; however all will agree that many of these injuries and deaths are preventable. Preventing firefighter injury and death starts with safety. It is the most important aspect of your job. Your well-being is your responsibility, so do not depend on someone else to keep you from getting injured or killed.

Many aspects of your job are uncontrollable. To give yourself the best possible chance of survival, you have to take advantage of the things that you can control. This starts with little things such as always wearing your seatbelt and maintaining a constant awareness of your surroundings. I nearly died at my first house fire because I was not aware of my surroundings. We arrived first on scene at a heavily involved residential house fire. I exited the engine and began pulling the number one cross lay off the engine. As I pulled the hose backward toward the house, a police officer frantically yelled at me to stop. When I stopped to look around, I was three feet away from a downed power line. My first fire could have easily been my last and it would have been my fault and no one else's. I should have been more aware.

You are issued protective equipment for a reason; wear all of it properly. This includes gloves and a face shield on EMS calls. Utilize your breathing apparatus and do not wait for the low air alarm to go off before exiting the building. Handle power tools in a safe manner. Always wear your eye and ear protection. Check for overhead obstructions before raising a ladder. Maintain accountability and never run on scene. Eat right and maintain your physical fitness. Heart attacks kill more firemen than all other causes combined. Get enough sleep and maintain your mental health as well. You are going to see a lot of tragedy in your career. You should never be embarrassed if you see something that bothers you. Chances are it bothered someone else too. Your department has someone you can talk with confidentially. Ask your officer, department chaplain, or human resources department about this service.

Continually gain knowledge. Do not stop at the classes offered by your department; ask them to send you to outside classes. Learn more about building construction, fire behavior, and scene size up. Learn to read smoke! Take advantage of every training opportunity available to you. When you are unsure of something, ask questions. There is no shame in asking a question that could save you or someone else from being injured or killed. The more educated you are about your profession, the greater your chances of avoiding pitfalls.

The highway is the most dangerous place on the job, because it seldom offers a second chance. People on the highway do not see you. They are either texting, intoxicated, or looking at the incident that you are working. For this reason, you should always try to have some type of barrier between you and them. This will usually be your apparatus. Make sure it is properly positioned and exit on the protected side if possible. Never rely on flares or cones. Wear your safety vest and keep an eye on traffic at all times.

Your job is inherently dangerous. It is perilous enough without you adding to the problem. You must look for every opportunity available and utilize every resource at your disposal to increase your safety. Do not depend on someone else to do this for you. Your safety is YOUR responsibility!

XIII. Wear Your SCBA

Your Self Contained Breathing Apparatus is the most important piece of equipment you have. It not only protects your lungs from smoke and super-heated gases that are waiting to sear them, but it also protects them from unseen hazards that will kill you just as quickly. There are other carcinogens and toxins that your SCBA keeps out. These toxins can slowly accumulate in your lungs over the course of your career and provide a slow painful death during retirement.

During rookie school you were likely drilled over and over on the proper care and maintenance of your SCBA. You were shown how to check it out each morning, how to ensure the proper seal of your face piece, and what the minimum acceptable cylinder pressure is before it should be changed out. Time was most likely spent in mazes and confined spaces, showing you its limitations and helping you become more comfortable with your SCBA. You were taught buddy breathing and emergency change out of a downed firefighter's SCBA. This was all done for a good reason. Your SCBA is your lifeline!

At Fires

Everyone knows you must wear your SCBA while fighting a structure fire. In fact, it is nearly impossible without it. The days of putting a wet rag in your mouth and going into a fire wearing hip boots are over. Because of the use of petroleum products in manufacturing, fires burn much hotter than they did thirty years ago. That coupled with the advances that have been made in Personal Protective Equipment, which allow us to travel much deeper into the fire, and you begin to realize that modern day firefighting could not be done without the use of an SCBA. The temptation to remove your SCBA will not come during the fire, but after. When the fire has been knocked down and a few windows have been opened, visibility will begin to return. At this point you will see some "old heads" start to remove their face pieces and turn their SCBAs off. They will talk to each other and work like it is no problem. As a rookie, you will likely think that you need to remove yours too, if nothing else to show them that you are a tough smoke eater as well. DO NOT DO THIS! I tried this once and damn near died, not to mention, I had a headache for a day. Even though the fire is out, the room still holds a tremendous amount of heat. The oxygen level is low and the CO level is through the roof. The air is full of things that will kill you twenty years from now. Wait until your safety officer gives you the all clear to work without your SCBA before removing it. This will usually be when CO levels have dropped below 30ppm. No one will think less of you for doing this and your retirement years will be much better.

Car fires and dumpster fires are two common calls where an SCBA is often neglected. You will often see senior firefighters not put on their face piece or not even wear their SCBA at all. This is a dangerous mistake and one that you must avoid. These fires are full of burning plastics, fuels, household chemicals and a ton of unknown carcinogenic substances. In fact, you never truly know what someone is hauling or what someone has thrown in a dumpster. To fight these types of fires, you must be in the smoke. There is no way to avoid it and you do not want to breathe that stuff. A car fire might also present the opportunity for a rescue. You would not want to live the rest of your life knowing someone potentially died because you were not ready to do your job.

Other Calls

There are several other types of calls that are routine responses where an SCBA is needed but often neglected. The first type is a gas leak or hazardous materials call. Gas leaks are routine and

On Scene

a majority of firefighters take them for granted. However, these calls kill or injure a large number of firefighters. One of my close friend's fathers was a 30-year fire captain in a large department, who made a routine gas leak call without an SCBA. He got too close and was overcome by the fumes. The result was CPR being performed on him. Hazardous materials calls are very dangerous as well. Many times you will be the first on scene and not know what type of substance you are dealing with. Some firefighters think that if they cannot see a gas then it is not there. Do not fall into this trap. Never approach a hazardous materials call without your SCBA on, even if other members of your crew do not wear theirs. They will need someone to drag them out when they collapse. The last type of call is the odor investigation or automatic alarm call. These calls are often in large warehouses. These calls are frequent and 99% of the time there is nothing to them, making them very easy to get lazy on. Most of the time, on this type of call, you will enter a building and find a clear environment. However, these buildings are very large and there could be a rapidly changing event taking place deep inside it. Once you encounter this event, there might not be time to make it back outside. You may only encounter this once in your career, but that one time could be life or death.

Remember that your SCBA is your most important piece of equipment. Know everything about it and maintain it meticulously. It is your lifeline. Wear your SCBA!

XIV. Smooth is Fast

I am sure you have heard the term "smooth is fast" at some point in your life. Elite combat units, emergency room groups, rescue squads, the fire service, and police SWAT teams use the phrase repeatedly. The common thread among all of these various organizations is that they operate in high stress environments, where seconds count.

I was a rookie firefighter when my eyes were opened to this harsh reality. My crew and I were working the scene of a two-story apartment fire with heavy fire conditions in two upper floor units and a portion of the common attic. We had various tasks on the fire ground such as forced entry, attack, extension and overhaul throughout the duration of the fire. Half way through the incident I was spent. My air tank was depleted and my body was exhausted. What was worse, when I looked around I noticed the other firefighters had plenty of air, stamina, and were still working. This is not a good feeling for a rookie firefighter. I changed my bottle, finished my work, but really was of no help to those around me.

After that incident, I did a lot of soul searching. I felt both embarrassed and alarmed about what had happened. I was alarmed because I realized at any point in a fire a life safety event may present itself. A situation could have arisen where I needed to carry a victim or downed firefighter, perhaps breech a wall to escape, or any number of other scenarios. Luckily none of this was needed, because at that point I did not have the strength or air to do it. I was embarrassed because I was young, in top physical condition, and had been out worked by firefighters who were older and had big guts. I could not figure this one out, so I posed the question to my captain. The wisdom he shared has benefitted me to this day. My captain said to me, "I know those guys aren't much to look at, but they deserve your respect." I knew that already, given they kicked my butt. He went on, "They know how to do something that you haven't figured out yet. They know how to make every step count." Once I heard this, it all became clear. These men could do the same amount of work I was doing, but with half the effort. They did this by staying calm, not making extra trips up and down stairs, having the right tool with them, making the tool do the work, knowing their building construction, and anticipating the next move. The majority of firefighters pick these tricks of the trade up slowly throughout their careers, however the sooner you can learn and apply them the better you will be.

From the moment the tone first goes off, take a deep breath and relax. Try to slow down the adrenaline rush. Adrenaline has a time and place, but it will also prematurely tax your body if you let it. On the ride to the scene stay calm and make sure your gear is on properly. Do not rush it. Mentally prepare yourself and take an inventory of the tools you may need for the job you might perform. When you arrive, take another deep breath and survey your surroundings. Do not get zeroed in on flames or carnage. You might get into a tight or scary situation, again take a deep breath and think. I was told in rookie school to expect the worse possible scene once you receive the call and hear the notes. This way, no matter how bad it actually is, it probably will not shock you and you will not have that moment of hesitation.

A good firefighter can look at a situation or series of events and anticipate what will happen next. You will know what needs to be done, what tools you will be need, where you need to be, and what hazards may be waiting to present themselves. In short, you will never be behind the curve or unprepared. Take these tips and add them to your toolbox. When you are able to begin applying them on scene, you will notice a difference. You will become more efficient. Keep practicing these principles and learning. Soon you will make every step count, and then you will know that smooth is fast.

XV. Do Not Freelance

You have probably heard it said many times before that "freelancing", "self-assignment" or "self-deployment" is not tolerated in the fire service. Engaging in this activity at an emergency scene is one of the quickest ways to piss off your crew, officer, and incident commander. Anger received for this activity is well deserved. Freelancing has killed and injured countless fire firefighters, as well as would be rescuers. It is in no way tolerated on the modern day emergency scene.

The Incident Command System is used by every fire department in the United States. ICS was developed back in the 1970's to control the chaos caused by freelancing. Before ICS, and even occasionally today, firefighters would do what they thought needed to be done or go do a job that looked fun to them. This usually resulted in most crews inside the house wandering around trying to get a piece of the fire, but nobody doing a search, salvaging any property, checking any utilities, or helping with overhaul. I have even seen freelancing result in two different holes being cut in opposite ends of a roof. The two crews were determined to start THEIR chainsaws and each were sure that the OTHER crew was performing it wrong and in the wrong location, even though just setting a fan would have worked all along! Worse than that, freelancing has resulted in firefighters going to perform a job alone and being trapped or killed, with no one realizing they were missing until it was too late.

Rookie firefighters fall prey to the temptation of freelancing for several reasons. Most think that they are just showing initiative or their willingness to work hard on scene. The best way to prevent this is to maintain strict company integrity. That means in a clear atmosphere firefighters stay in visual and voice range of each other. In an Immediately Dangerous to Life or Health (IDLH) atmosphere they must be able to physically touch each other.

As a rookie on an emergency scene you should stay in arm's length of your officer or crewmembers. Stay in their hip pocket and be able to grab their coattail at any time! No one wants to have to look around for you. There are usually only three exceptions to this rule. First, if you are

The Station Ready Rookie

on the first arriving crew and you are assigned a task such as deploying attack line while your officer does a 360 size up or gets crucial information from bystanders. Second, your officer goes to have a face to face with the incident commander. Third, you are told to go retrieve equipment from the apparatus. In the first two scenarios, it is your job to know exactly where your officer wants you to wait for him or her. This will usually be in the front yard or at the front entrance of the structure; it should NEVER be in an IDLH atmosphere. In the third scenario, your job is to go directly to the equipment and directly back to your officer or crew without deviation. Their job is to be in exactly the same place as you left them. If for some reason you return and they are not where you left them, contact them on the radio and find their exact location. Remember, entry into an IDLH atmosphere requires a minimum of TWO people. You should never be alone. If your officer assigns you to help another crew, maintain those same standards with your newly assigned crew.

Remember, a good reputation on the fire ground or emergency scene can be earned while adhering to these rules. Your reputation will be earned by having the right tool, knowing how to properly perform the assigned task, your attitude, your physical conditioning and a willingness to work hard. Not adhering to these rules will earn a trip from the chief and the chaplain to your door. A trip no one in your family wants.

XVI. The Right Tool for the Job

Knowing the proper tools needed on scene, and having all of them with you, is a critical task for any firefighter, especially a rookie. Tool selection and having those tools ready, working and available will often make the critical difference in the extent of property loss or life safety. Having to be told to get something that you should have already had in your hand and making extra trips to the apparatus wastes costly seconds. Remember that an emergency scene is ever evolving and usually grows more dangerous by the minute. The scene will be hectic and everyone on your crew has a job to do. That job is not babysitting you! At the very least, knowledge of tools and their selection will make you an asset to your crew.

On Scene

Tool selection will differ slightly by department and geographical regions of the country. Tool preference can sometimes even differ between various officers and shifts within a certain department. The best thing to do is to sit down with your officer and find out exactly what he or she expects. The station officer will usually try to explain to a rookie what is needed in certain situations. Sometimes this is done on the first day; sometimes it is done in route to a call. It is impossible however to cover every situation or think of every emergency. In special cases, the right tool will not be known until your officer sizes up the situation. However, there are some general rules that you need to know for riding different apparatus, just in case your officer does not tell you his or her preference, or you swing to a new station and ride with a different officer.

Early in my firefighting career, my engine company pulled up to a heavily involved residential structure fire. I exited the engine empty handed, deployed and readied my attack line. When the line was charged, I kicked in the front door and proceeded to attack the fire. When the fire was knocked down, my officer and I vacated the structure. I was expecting to get a high five for what I viewed as a "kick-ass" job. Much to my surprise, my officer scolded me for not having the right tool in my hand and using it to force the door. He brought to my attention that by kicking the door, I ran the risk of blowing out my knee. Had I blown my knee out, the entire first in attack crew would have been lost to my injury. Tools are there for YOUR safety as well. After that, I never made the mistake of stepping off an apparatus empty handed.

On Your Person

You will need to carry various tools in the pockets of your bunker gear at all times. These include but are not limited to: cable cutters (not linesmen pliers), multi-tool, a short section of webbing (15 feet), small flashlight, pen, paper, foam ear plugs, small shove knife, two small wedges, small diameter aluminum spanner wrench and voltage tester. This sounds like a lot, but each will be needed at some time. This will also add some weight to your gear, so do some squats.

Cable cutters will be needed to cut the battery cables at the scene of car wrecks and to possibly cut you loose from entanglements on a fire scene. Be able to find and operate them with a gloved hand. They might save your life. The multi-tool is a good way to carry pliers, a knife, and screwdrivers in a small package. Webbing is great for multiuse. It can be used as a tag line for searching small rooms, used to secure ladders or tools, used to repel in a Mayday situation, or used to drag a downed victim to safety. The small flashlight acts as a backup to your large flashlight and aids in morning check out of the apparatus. Pen and paper are self-explanatory. Remember to buy a waterproof tablet or to put yours in a plastic bag. Foam earplugs are handy for fire alarm calls or fire drills. Most old firefighters are deaf, so save your hearing now. A small shove knife will often gain you entry into a residence when the occupant cannot make it to the door. They are small and do minimal damage. Your wedges will be used to prop a door open so that the attack line doesn't drag it shut; or to stop the flow of an open sprinkler head. The small spanner is lighter than a crescent wrench and has a slot in the end for shutting off gas and water meters. The spanner will come in handy during hose break down and can be used to wedge a door open. A voltage tester can be put into an electrical outlet or laid on an electrical cord to see if it is energized. This adds safety when overhauling a fire scene or working around machinery. They are small and inexpensive.

Engine Tools

As you know, task and tool selection will differ by apparatus. A fire engine's main job is to apply water to a fire. With that being said, your first tool to know how to select and deploy is fire hose. Your engine will carry an assortment of supply and attack hose in different lengths and diameters. You must know how to select, deploy, couple and reload all of them. It is best to know your officer's preference beforehand. If you are deploying supply line, remember to always have a plug wrench, spanner wrenches (2), and FDC key. Most departments bundle these in a "hydrant bag".

If you are not the firefighter deploying the line, you will need a tool in your hand. This should be a set of irons, or a 4-foot drywall rake. Irons are great multiuse tools and can cover an array of tasks on the fire ground. They should also be with you when you approach a vehicle collision as well. You will never be wrong stepping off your engine with a set of irons in your hand. However, they are heavy and bulky to carry. A great alternative at a fire scene is the 4-foot drywall rake. It is light and can be carried in your belt if you are the nozzle man. This tool is also great for quickly checking extension and overhauling an area after the fire has been knocked down. The drywall rake is also a good tool for extending your reach while sweeping and searching a room. Along with these hand tools be sure to have a large flashlight, a radio and thermal imager on your person if possible.

Truck Tools

An entire book can and has been written about tools carried on a truck and the selection of such. So this will be kept simple. Like the engine, you will never be wrong stepping off a truck with a set of irons in your hand, a radio and a large flashlight on your person. Truck companies are assigned a variety of tasks on an emergency scene. When choosing your secondary tool, you must listen for the assignment your truck has been given. You should know from rookie school whether the assignment calls for salvage tarps or a chainsaw. The main thing with apparatus tools is to memorize where they are located, be an expert in their use, and meticulously maintain them. No

On Scene

matter what, do not step off an apparatus empty handed. Always have a tool in your hand.

XVII. Fight Fire Like You Have Been Trained

The fire ground is a very dangerous and dynamic atmosphere. One small misstep at any time can catapult the whole scene out of control. When this happens, the chance for firefighter injuries or death multiplies exponentially. To combat this potential, you must use the tried and true firefighting tactics that you have been taught. You must collect yourself and remember your fundamentals. We will review some of those below.

If a scenario is going to end well, it usually must begin well. This means the first arriving engine must lay the groundwork for a successful outcome. You must select the correct size and length of hose as well as properly deploy it.

Rushing this step has the potential to lead to many problems later on. You run the risk of your hose becoming tangled, robbing you of pressure and making it unable to advance inside the structure. Selecting the wrong length or diameter of hose will make you come up short before you reach the seat of the fire or not give you adequate volume for the fire load. Either way, the fire continues to burn and attack the structure, making it more dangerous for everyone on scene.

Once your hose line is properly deployed, you must choose your plan of attack. The most overlooked action is that of protecting exposures. Firefighters are drawn like "moths to a flame" and forget what is most important. Do not zero in on a fully involved structure when you could save someone's property next door. Protecting an exposure is not fun or glorious, but it one of the most important jobs you can do. If fighting the fire is the correct course of action, do not just run in a structure blindly. First you need to make a mental note of the egress points. Size the structure up and read your smoke conditions. Have an idea in your mind of the interior layout of the structure and the location of the seat of the fire. This is best accomplished by talking with a resident who may be standing at the scene. Spending thirty seconds gathering information could make a lifetime of difference. Check your nozzle pattern. Once you are prepared, attack the fire from the uninvolved area of the structure.

When entering a structure, check for fire above your head. Open concealed spaces. Many firefighters have walked into a clear building and later died because fire was running unchecked above them the entire time. Had these firefighters opened the ceiling at the beginning, the outcome might have been different. Never pass fire. Once you are in the structure, stay low. Even though standing up might not physically burn you, the hotter temperatures are extremely taxing on your body. Staying below these as much as possible will help to better conserve your energy and will give you the best visibility. Often when a room has heavy smoke conditions, there is still some visibility at the floor level. Get as low to the floor as possible and take a good look around. This will give you a snap shot of the room. Keep your reference points and never stray off the hose line. Be alert and listen for any sounds within the structure. You may be able to hear a victim or the sound of the fire burning. Utilize all of your senses.

Once you arrive at the fire room, take a moment to survey the area before you apply water. This will give you a new set of reference points to work from. Especially note where the windows are. This will allow you to move to them and ventilate the room more easily. Once you black the room out, the chance to get a visual is lost. Utilize proper nozzle patterns and techniques when applying water the fire. Avoid disrupting the thermal balance of the room as much as possible. After the fire has been knocked down, move to a

On Scene

window and hydraulically ventilate the room. This is a highly effective and underappreciated method of ventilation. Hydraulic ventilation will quickly clear the room of smoke, allowing you to check for victims or fire extension.

An emergency scene often holds too many variables to list. To control those variables as much as possible, you must use the proper techniques that you have been taught to fight fire. Remember the basics and use your senses. Rushing the process and missing a crucial step will likely put you and many other people at a greater risk than necessary.

XVIII. Learn Building Construction

Great warriors study their adversaries and battlegrounds thoroughly. They are familiar with their opponent's strengths and weaknesses and know how to best utilize the playing field. The same battles are waged on the fire ground. You are trying to keep a building up, and fire is trying to bring the building down. In order to be successful, you must know the strengths, weaknesses, and limitations of the building in which you will wage war.

As a rookie firefighter, you need to be familiar with the five types of building construction. You must also be familiar with common building materials and their level of fire resistance. If you are not well versed, consider revisiting the subject in depth. Building types and methods of construction will vary for different time periods and areas of the country. Everyone will not have high-rise structures or homes with basements in their fire district. Your goal should be to recognize the building types and methods of construction used in your area.

Start by surveying the size, type, and ages of the buildings in your area. They all have a lifespan. Determine the age of your district. Fighting fire in a two-year-old house is very different from fighting fire in a sixty-year-old house. The structural systems of newer buildings often fail much faster than those of older ones. Have the buildings been remodeled, concealing their original construction type? A number of occupancies have been added on to, using different types of materials than those used in the original structure. This often makes one area of the building safer to fight fire in than the other. Notice if there are metal stars or anchors on the outside walls, signifying that it has been reinforced due to structural problems. Pay particular attention to roof construction. Roof collapse has claimed the lives of many firefighters. Look for buildings with large open spans, such as churches or assembly halls. Stop by construction sites and see the types of roof trusses being used in your area. Wood trusses with gusset plates and unprotected open web trusses fail extremely fast when exposed to fire.

Knowledge of building construction is vital in the fire service. We deal with it on a daily basis. You must make it a priority to learn as much a possible about building construction and the structures in your area. Become familiar with different types of materials and their level of fire resistance. Study the battlefield.

XIX. Learn to Read Smoke

The ability to read smoke is one of the best tools you can add to your toolbox. It gives you the ability to make a rapid assessment of fire conditions as you approach an incident. This assessment is a key to your success and survival on the fire ground. When read properly, smoke will answer four questions: location of the fire, size of the fire, where the fire is going-and how fast, and the potential for a hostile event.

Reading smoke starts by looking at four key factors: volume, velocity, density, and color. In general, the volume is representative of the amount of fuel burning. Velocity is indicative of pressure buildup. It is caused by either heat or volume. Density is the most important safety factor in reading smoke. The thicker the smoke, the more dangerous it is. Color can show the material burning, the stage of heating, and/or the distance from the fire. Very black smoke is highly explosive. Smoke will generally be darker closer to the seat of the fire. Turbulent smoke is caused by high heat conditions and is an indicator of flashover. Smoke that momentarily exits, but is pulled back inside shows a likely backdraft.

Many things influence these four factors. The first influence is the size of the structure that the smoke is coming from. Secondly, weather will affect smoke behavior. Thirdly, firefighting efforts will influence smoke behavior. All four factors of smoke should change during firefighting operations. The volume should initially increase, the velocity and density should decrease, and the color will change to white.

Learning to read smoke could very well save your life one day. It is a tool you must possess. The smoke will tell you if conditions are getting better or worse. When correctly done, it will give you the size, location, and speed of the fire. More importantly, it will predict a catastrophic event that could potentially injure or kill firefighters. This is only a brief overview of the subject. For further study, you should find material from Dave Dodson. He is the subject matter expert.

XX. Get Ready to Drive

Driving an emergency vehicle is a privilege few people will ever know. Though often overlooked, it is one of the most important jobs in the fire service. In fact, unless an emergency happens at the fire station, it is involved in every call we make. The driver is charged with knowing the proper route to get to the destination, safely navigating the roadways, properly positioning the apparatus once on scene, and effectively operating it as needed. This is a huge responsibility.

Many departments do not allow rookie firefighters to drive fire department vehicles during a code three response until they have at least one year of service. If you are not there yet, do not worry. It will we be here soon enough.

On Scene

There is much you can do to make sure you are ready when that day finally arrives.

Start studying now. You will need to learn every street in your district, the location of FDC connections in your district, hydrant locations, major streets in your city, as well as routes to all hospitals that your ambulance may use. This is a daunting task, but one that can and must be done. Keep a map rolled up in your back pocket at all times; review it during any down time and on your days off. Also keep in mind that knowing a street on paper is not the same as being able to drive to it during an emergency situation. Pay attention when you are going to and from calls. Learn the landmarks, blind spots, and special hazards of your district. When you are relieved from duty, spend an hour every morning driving your district in your personal vehicle. Study your apparatus. Help the driver during morning checkout, and learn his or her routine. Know its specifications and capabilities. Ask your driver or a senior firefighter to spend time familiarizing you with the apparatus. Learn to operate it, and every tool on it, safely and efficiently. You want to be the best driver you can be. When the call comes at 0300, you will be in a daze and full of adrenaline. It is at this time when your study will pay off and enable you to make it on scene safely.

There are several considerations for operating the apparatus. Be smooth. Smoothly accelerate and decelerate the apparatus. Your crew will appreciate this, especially if you are driving an ambulance crew to the hospital. Maintain your speed. When running code, a safe speed is usually 5mph above the posted speed limit on city streets and 10mph above the posted speed limit on highways. Exceeding these limits puts your crew and the public at an unacceptable risk. Stop at all intersections and stop signs. Your lights and sirens only ask permission; they do not give you the right to blow through red lights. Drive defensively. Be alert to your surroundings and always have a backup plan. Wear your seatbelt and make sure your crew is wearing theirs. Following these guidelines will help you and your crew to arrive safety on scene. Not following these guidelines could be disastrous.

Learn where your officer wants the apparatus positioned on different types of calls. On medical calls the front door is usually left for the ambulance, with the engine or truck parked behind. This shields the ambulance crew while loading a patient. Remember to leave them plenty of room to operate. At vehicle accidents, the engine or truck is parked behind the wreck to shield those working from oncoming traffic. Park your apparatus angling away from the accident with your wheels turned. This allows you block an extra lane, as well as deflect a glancing blow from a vehicle and not have your apparatus pushed into the accident scene. Never park just over a hill. Make sure traffic can see you. Utilize your cones, flares, and emergency lights. When arriving on the scene of a residential structure fire in an engine, pull just past the house. This gives your officer a three-sided view for his or her size up and leaves the address (the front of the structure) for the truck. At commercial fires, park your apparatus out of the collapse zone. If you are driving a truck, spot it on a corner of the building that is free of overhead obstructions. These are just a few common scenarios, but they will give you a good foundation.

While we have only scratched the surface of what it takes to be a good driver, you should have an idea of what is needed. Study your district and city. Learn to operate your apparatus and to position it properly. Continually improve on all aspects, as it will not come overnight. However, with hard work and dedication, your officer and crew will be proud to ride with you. Drive safe!

Section 3: Attitude Is Everything

XXI. Take Pride in Your New Career

First and foremost, congratulations. Whether you are paid or volunteer, big city or rural, certified or headed to rookie school, you are now a firefighter. This is one of the most trusted and respected professions in the world. Being able to tell people that you are a firefighter carries a lot of weight, and a lot of responsibility. As far as your friends and family go, they considered you to be a firefighter the day you got hired. However, as you may have guessed, the men and women that you will be sharing a station with want to see a little more out of you before they are ready to fully accept you into the group. The previous sections will go a long way towards getting you there, but in the end most of the opinions about you will come down to your attitude. That is why the section is titled "Attitude Is Everything".

The very foundation of that attitude starts with the pride that you take in being a firefighter. To your friends in other professions, that might mean running into burning buildings or cutting someone out of a wrecked car. But, as you have learned from this book already, it is not all glamorous rescues. You will have to take pride in every aspect of your new job, from scrubbing the toilets to cooking dinner; because to the people that you have to impress, your new crew, those are your most important jobs. If they cannot trust you to perform the small tasks correctly, they will not have any confidence in you to complete assignments in an emergency scenario.

Your family at home is already tremendously proud of you. They are telling all of their friends how their son or daughter, husband or wife, is now a firefighter. They are going to put pictures of you in your new uniform all over the house. But, your new family in the fire department wants to see what kind of pride you really have in the job. Are you willing to do the dirty work? Will you take their jokes? Do you know what you are doing? They will be much harder to impress than your actual family, but once you win them over you will be a member for life. Just remember to take pride in your new career, and all that it entails. You might have gotten the job, but you still have to earn the badge.

XXII. Learn the Rules

Every fire department has two sets of rules: formal (written) and traditional (unwritten). You will need to know both. The formal rules should be fairly easy to find and to follow. Your department will have a catalogue of standard operating procedures or guidelines (SOPs or SOGs) and/or directives. There will also be a protocol book for medical calls. You need to learn all of these policies. They are the basis for how your new fire department operates. If you break any of these rules, ignorance will not be a valid defense. Whenever you have time, go over these formal directions until you are familiar with all of them. This will be a big help to you on scene, and throughout your career.

Attitude is Everything

Every fire department also has a unique set of procedures that you will not find written down anywhere. Unwritten rules, or traditions, have evolved over time. Many have been around long before anyone that you will be working with. While they are certainly more difficult to find, they are every bit as important as the written policies. The rules vary so much from department to department that it will not be much use to list any here. Traditions can vary greatly from station to station, or shift to shift, within a department. The best way to learn unwritten rules is to ask and to be observant. There should be one person who mentors you, usually the next youngest firefighter at the station. This will vary between departments and might be one of the unwritten rules. He or she is generally the one that you should ask about these types of things. As the rookie you will be expected to break a few of these rules. When it happens, keep a good attitude, apologize, and remember the rule so that you will not break it again. Do not criticize the tradition or try to change it.

One example of an unwritten rule is where people sit at the dinner table. Everybody has his or her regular spot. One of the favorite jokes for firefighters to play on the rookie at dinnertime is to try to get him or her to sit in the officer's seat. This should not happen, because you should be the last to sit down. But occasionally someone will be late to a meal because they are on a run. This is exactly what happened to me. I swung to a different station, where I had never worked before.

The truck crew was gone on a MVC call. So, when the food was ready, and everyone who was at the station sat to eat, I asked where my seat was. I sat where I was told. When the truck got back to eat everyone had a good laugh because I was in the captain's seat. He took it pretty well, but it was enough to embarrass me. I made sure I always knew where everyone's seat was from then on.

XXIII. Respect Your Elders

In the fire department, seniority is everything. You will quickly find that there were no decent firefighters hired after whomever you are talking to. He or she was in the last great class, the last of the real firefighters. It does not matter if they were hired 30 years ago or 3 years ago. This is the way firefighters view things. Whether or not they were the last of the real smoke-eaters is certainly debatable. But what is not debatable is the fact that they have more experience than you do. Experience as a firefighter is invaluable. All the classroom knowledge and theory about firefighting is useless until you actually have a nozzle in your hand. Venting the roof at the training field is not the same as venting the roof on a warehouse fire in the middle of the night. You will hear the same stories repeated many times.

Attitude is Everything

People will tell you about how tough things were in the old days, and how great you have it today. You will hear about the one fire that can never be topped. You should sit and listen intently to all of them. This is one of the best ways to show respect to the people that have paved the way for you to have this job.

Address firefighters by their title (Chief, Captain, Lieutenant, etc.). If they do not have rank, Mr. or Ms. will do. When they are ready for you to address them by their name, they will tell you. I would still use the formal title one more time just to be sure. However, if they are uncomfortable with you doing this, then obviously do not continue.

The captain from the previous shift would always be up early drinking coffee as I was cleaning in the morning. I would always say good morning, and address him as captain. He told me once to call him by his first name. I kept addressing him as captain, until my rookie year was over. That next morning, he took me aside and told me now that I was not a rookie anymore I was going to address him by name. I know he appreciated the respect that I showed him, and now he wanted me to know that I was part of the station. This was an individual case. Remember, that the fire service is a para-military organization and that a senior member should be addressed by their title or rank, unless they say otherwise.

XXIV. Listen More Than You Talk

Do you know the old saying about why you have two ears and one mouth? The answer, of course, is to listen twice as much as you talk. That advice will serve you well in all walks of life. It is especially important in your first year at the fire station. However, we will focus on the fine line between talking too much and being aloof. Your new family will want to know who is in their house, but they will not want to hear about everything that you know or how great you are. If you are too quiet and keep to yourself, they will think that you do not want to know them. On the other hand, when you start giving up too much information; you will hear about it later. Just try to relax and be yourself.

Attitude is Everything

One of the most important things that you can do to make a good first impression on everyone is to approach him or her, shake their hand, and introduce yourself. Do this to every new face that you see. If they want to know about you, they will ask questions. Others may want you to prove yourself first, and that is fine too. You will have plenty of opportunities to show your character and work ethic.

One of the best ways to get to know your crew is one-on-one. Everybody has their specialty, and they will be happy to teach you what they know. These teaching sessions are a good time to build personal relationships. Keep in mind that no matter how much you have done before or how well you did at the fire academy; you know nothing. You are the rookie, and you are expected to learn not teach. If you learned something a different way than what you are being shown, it will benefit you greatly to do it the same way that the rest of your crew does. You will have opportunities to show off your skills later. For now, just listen and do as you are told.

There is a guy that I went to rookie school with who can go days without saying a word. He had a rough start at the station. His crew thought he did not want to be there, or did not care to be a part of their group. He finally opened up and started talking to them and they really loved him. The problem was, everyone else in the department still had the same opinion of him because they had not heard him say more than two or three words.

The lesson is to stay away from either extreme. Do not be a loudmouth, but do not be isolated either.

XXV. Do Not Be Afraid to Ask Questions

It is certainly better to ask what is expected of you and get teased for asking, then to have no idea what is going on and have everyone talking about how you do not know what you are doing. You should sit down with your new officer on the first day and request what is expected of you. He or she will probably do this anyway, but they will also appreciate you taking the initiative to ask first. You should find out everything expected of you around the station, and on scene. Some officers have standing orders, some treat every incident differently, and some just want you to stay by their (or someone else's) side. It is critical that you find out what your role is. Also, make sure you inquire about any unfamiliar equipment. Your crew will be happy to teach you how to everything works. Because when the time comes, whether you have asked about it or not, you will be expected to know how to use it.

Another good question to bring up is: how am I doing? Believe me, you will want to know. But for some reason, most new rookies are scared to ask. This should be done in private, with your mentor or officer, not at the dinner table. You should be looking for honest feedback and not a pat on the back. Open and honest communication will benefit you and your whole crew. If you have any weaknesses, it is best to get them out in the open. It might be something that you are not even aware of. But, once it is out in the open, you can work to correct it. If you have a weakness that you do know about, it will be exposed no matter how much you try to hide it. So again, get it out in the open and start working to correct it. There is no shame in asking a question. Everyone had to learn at some point in time. Firefighters will be glad to teach you anything they can. However, if you do not have any questions, then they assume you know the task at hand. The only stupid question is the one not asked. Never be afraid to ask a question, and never have to ask the same question twice!

XXVI. Learn From Your Mistakes

You will make mistakes. The rookie is expected to make errors. If you are not making any mistakes, then you are not trying hard enough. The key is to not make the same misstep twice, and to never make excuses. When you have a slip in judgment, listen to the person correcting you and know why you are accountable. After they explain what you did wrong, confirm the correct procedure, make a mental note, and move on. Never voice an excuse, even if it is valid. Your crew might give you a hard time, but they know it is part of the process. At some point, you will hear stories about all of the rookie mistakes they made. This is an important custom in the fire department; you can learn a great deal from the mistakes of others. When you fall down, get back up.

Attitude is Everything

Whatever you do, never give up on a task. This is one of the things that your crew will be watching closely to see if you reveal any clues to your character. If you are having trouble with a task do you get frustrated and quit, or do you strap in and get the job done? There will be times at an emergency scene when things are not going your way, conditions will be terrible, and nothing will seem to work. At that point, you will not have the option to get frustrated and walk away. The job has to be done, and your crew is counting on you. They want to see what you will do around the station before you get to a real emergency situation. Earn their trust by never giving up on a task. Find a way to get it done.

XXVII. Train

Training is everything. In today's fire department you just will not get enough experience to learn all that you need to know. Thanks to modern building codes, structures do not catch on fire as much as they used to. Some cities still have a good amount of fires, but they are the exception rather than the rule. You have to be prepared for any one of a million different situations that might occur when you show up for work in the morning. Some of which you might only experience once in your whole career. How do you prepare for this? You train. You know your equipment, your responsibilities, your crew, your standard operating procedures, and good technique. You will certainly have mandatory and coordinated training with your station and shift, but you should always be drilling. When you are

Attitude is Everything

not doing station work or on a call, you should have a tool in your hand. Learn all about the power tools, practice your knots, throw ladders, fold tarps, and learn your apparatus.

Any down time should be spent preparing. This includes studying. As a rookie, you will have scheduled evaluations over a variety of topics to make sure that you know what you are doing. You are expected to be prepared for these, and not just pass, but excel. This will take studying. I know most people who take this job do not do it to have their face buried in a book, but you should never stop learning. Believe it or not, all of those things that you are being tested on can be used on emergency scenes. There is no excuse for not knowing how to do something that could have easily been learned by opening a book for five minutes.

For some reason, scheduled training days are some of the highest sick days of the year. I think people are embarrassed of making a mistake, or not knowing something. That is the exact point of practice, to learn what you do not know, and to fix it in a controlled environment. It is not hard to see who skips training days on the fire ground. They are making the mistakes that they should have made during a drill. Not only does everybody see his or her blunder, it is magnified. Develop good training habits early and seek outside education opportunities.

XXVIII. Actions Speak Louder Than Words

As a new rookie, you will have to earn your respect. No matter how much you know, or think you know, this will never be done with words. It is all about actions. You might be able to talk a good game, but all that counts is if you can back it up. The sooner you can show, not tell, everyone your character and commitment, the sooner you will be accepted. You should be self-motivated and always looking for something to do. Show enthusiasm and take pride in your work. There is no need to tell everybody how great you were at the academy, or your last job; nobody cares. If you worked for another fire department, it is best not to mention how you did things there. Every organization takes a lot of pride in the traditions that it has built and the way that it operates.

Always take action, but do not be a showoff. Remember, this is the ultimate team atmosphere. Everything revolves around your crew. If you are better than everyone at something, then it is your job to stay humble and to help out. People will notice your abilities without you having to tell them. If you boast, all they will remember will be your bragging and not your talent. You do not want to be resented for being good at something.

Volunteer for everything. You already know that all of the housework and cooking is your responsibility, so that is covered, but there will be other things that come up that you will be expected to handle as well. You can volunteer and look like a hard working, motivated rookie; or you can wait to be told that you have to do it and let everyone second-guess your commitment. There will be demonstrations, drills, station tours, school visits, neighborhood events, union activities, and many other chances to show your enthusiasm. It would be especially beneficial for you to volunteer for things that take place outside of your normal shift. This will really give you a good name, and give everyone a chance to meet you. While you might still be treated as a rookie, outside events will be much more relaxed.

My first day at the station, I went to introduce myself to the driver of the truck. I could hear some giggling as I walked up to him, but was not sure why. As I started to introduce myself, he cut me off and told me that he did not care who I was (in much rougher language). This guy was notorious in the fire department for being hard on everybody, especially rookies. Well, later that day we got 1200 feet of new 5 inch supply line for his truck. I was the first one out the door to load it, and the last to go inside. It took some serious work and a lot of sweat to load in the 107-degree heat. After that, he bought my dinner and started talking to me. We are still friends today, and he has taught me as much about how to operate on the fire ground as anyone. He just wanted to see me prove myself before he took the time to get to know me. Do not be discouraged if someone reacts to you the same way. Just stay respectful and get to work.

XXIX. Watch Out for 10/20 Syndrome

This can be phrased many different ways, but the point is the same. Do not get 10 months (or any small amount of time) on the job and act like you have been there for 20 years. You are at the bottom of the pyramid, and are expected to act like it either until your year is up or there is a new rookie below you. At some point early in your rookie year, your crew will try to get you to relax with them and watch some TV or play video games. This is a trap; do not fall for it. They want to see if you will act like you deserve to be there before you have proven yourself, or if you have 10/20 syndrome. You should be busy working or studying.

Attitude is Everything

However, there are times when your crew might genuinely want you to take a break, or just take some down time and get to know you. If this is the case, they will make you sit down and relax with them. Take advantage of the time to bond with your crew, but make sure that all of your work is done and that you have not overlooked anything. They might genuinely want you to watch the big game with them, but they also expect the toilets to be clean and the dishes to be washed.

One morning, after my first year, when I came in for work I sat down on the couch with the rest of my crew. On the way out the door, one of the firefighters from last shift said bye, but he called me by the wrong name. He called me the name of their rookie. I thought he was trying to take a shot at me, so I asked him what he meant. He said their rookie plopped down on the couch and did not move all day, so he just figured it was still him in that same spot. Everybody got a good laugh, but that rookie had started a reputation for himself that will be hard to shake.

XXX. Leave Your Opinions at Home

By now you should know your place as a rookie in the fire station. You are entitled to your opinion, but you do not necessarily have to share it. Once you have proven yourself, and are a trusted member of your new crew, your opinion will be valued. Now is not the time. Much of what you have learned, or will learn, at the fire academy is different than what your crew was taught. As much as the fire service loves tradition, there is still constant innovation. You have just been taught the latest and greatest way of doing things. Your crew however, has their own way of doing things that may be a little different from what you have learned. They have taken what they were taught at the academy and expanded it through years of experience. They do not want to hear the latest book method that is fresh in your mind. Many new rookies come out of school full of excitement and enthusiasm for the job and want to teach their department the new techniques that they have learned. You should wait until you have some time on the department, then you will be able to share your knowledge.

One thing to watch out for is being tricked into saying something that you should not. Some people will try to get you to voice a negative opinion so they can embarrass you with it later. They want you to say something bad about the old driver, or the lieutenant, or whoever, so they can bring it up at the dinner table and have a good laugh at your expense. If it is something that you really said, it will be hard to talk your way out of. Obviously, the easiest way to avoid this is to keep your opinions to yourself for now.

Gossip can get you into even more trouble than an outrageous opinion. The fire department has more gossip than any sewing circle, teachers' lounge, barbershop or high school that you have ever seen. It will be a great benefit for you to avoid it. Any rumor or gossip that you pass along can, and usually will, come back to haunt you.

It is also important that you make up your own mind about everyone in your department. At some point, you will want the same from everyone else. Get to know each firefighter, and form your opinions based on what you see personally. Past history plays a big role in the fire department.

Firefighters can hold grudges against each other dating back to rookie school, past stations, or maybe even the town they came from. You will find that some people live up to their reputations, good or bad, and some do not. I have been amazed at how many people that I have heard bad things about and turned out to be really good on scene and around the station. You will find that more reputations are based on personal feelings and grudges than actual performance.

XXXI. No Temper Tantrums

As a rookie you will be tested to see what you are made of. You will be picked on, teased, and probably made to feel uncomfortable at times. This is all a test to see how you react. Most of the lessons that are taught will be at your expense. This is all part of the process. Make sure you keep a good attitude and a sense of humor. Sometimes it will seem like nothing you do is good enough. Your crew will find something to nitpick. This is another test. They want to see if you will give up, quit, or throw a temper tantrum. Do not let it get to you. Keep working hard and staying busy. If you are truly doing something wrong, they will correct you. When in doubt, you can always ask.

If however, they see that they are getting to you, the harassment will only get worse.

Attitude is Everything

Firefighters love to get each other wound up. If a joke is made about you, laugh at it and do not take it personally. If someone sees that you are an easy target, they will be relentless. Word of this will also spread to other stations. Do not add any fuel to the fire, and do not take it personally. You might get more that your fair share, but you will soon see that everyone gets their turn. Sooner or later, you will have your own rookie and you will have to decide how far to push them.

XXXII. Learn to Love EMS

The majority of people who fill out all of the applications, the background information, and wait in long lines for hours to test, only to compete against hundreds or thousands of other candidates, before going through a grueling hiring process, just to work tirelessly in rookie school, do so to fight fire. That is just what most people think of when they picture what we do. The modern reality is much different. Whether or not your fire department provides ambulance service, or it is provided by an outside agency, the overwhelming majority of your calls will be EMS related. That is just how it is, so you better learn to love it. Some of you might already, make sure you use that enthusiasm to excel.

Attitude is Everything

    Some of you might have been hired because of your Paramedic certification or experience. You will be expected to jump right in and provide patient care. For new rookie EMTs, you should be ready to help the medics carry all of their gear, get the stretcher, and provide all of the patient care that your certification allows. This might just mean getting vital signs and moving the patient to the cot. The more you can do to help out, the better. In fact, you will probably have an easier and more visible time making a contribution on medical runs than you will at a fire. Use this to your advantage. Show that you work proactively on scene and always anticipate what needs to be done.

    Treat people with respect and dignity. Some members of your crew may be short or rude to patients. Do not let yourself fall into this rut. It does not matter if you deem the situation an emergency or not. The patient made that decision for you when they dialed 911, now it is your job to help. When I went through paramedic school, they taught us to use the "grandma test". Treat every patient the way that you would want someone to treat your grandmother if she needed help. This standard of care will be noticed immediately.

Conclusion

Many hard lessons from first years in the fire service are condensed into the preceding pages. The strength of any fire department, large or small, lies in its personnel. Regardless of what part of the country they are located; firefighters are a breed of their own. Pride, tradition, and loyalty are but a few of their universal traits.

The fire service is often referred to as 100 years of tradition uninterrupted by progress. We do not believe that to be the case. It is our hope to make progress in this profession starting on the front end. This is the most overwhelming and life-changing time for any firefighter, yet it is often the most overlooked. Firefighters need to be given the tools to both survive and thrive, starting with their first day on the job. We often train for events that we will never encounter and somehow overlook

Conclusion

the small lessons that will set the tone for a firefighter's entire career. Once the damage is done, it is very hard to undo. We must start the new firefighter off on the right foot and give them every opportunity for success.

We believe this book will help to effectively maximize any department's current personnel, any training academies future recruit, or any individual firefighter who utilizes the information contained within. By applying the principles, the reader will be well on their way to being the best firefighter they can be, and strengthening any department that they belong to.

Lastly, we know this is a serious book with lots of rules and warnings. But always remember that one of the most important parts of this job is FUN. This is one of the greatest careers in the world not only because of the reward of helping other people, but also because you can have a great time doing it. Make sure that you enjoy yourself. Good luck!

Feedback

We sincerely thank you for taking the time to view this material. We constantly strive to make our rookie manual more relevant and beneficial to the end user and the fire service in general. We welcome and appreciate any feedback that you may have. Please let us know if there is something that we missed or something in particular that you feel is needed to make future firefighters' rookie period more enjoyable. We value your thoughts and opinions.

Please send all feedback to: info@rookiehelp.com

About the Authors

All three authors work for a major suburb near Dallas, Texas with a population of approximately 230,000, making it the 12th most populous city in Texas and the 87th most populous city in the U.S. The fire department employs almost 250 personnel handling over 20,000 Fire/EMS calls per year with a budget of $21 million.

Captain Collin S. Blasingame was hired in 1997. He promoted to Driver/Engineer in 2002, to Lieutenant in 2007, and to Captain in 2010. Captain Blasingame holds a Bachelor of Arts in Public Safety/ Administration from Stephen F. Austin State University. He also served as Director of Emergency Services for his hometown.

Lieutenant Justin C. Dickstein was hired in 2006. He promoted to Driver/Engineer in 2008, and to Lieutenant in 2012. Lieutenant Dickstein holds a Bachelor of Arts in Government and a Master of Science in Technology Commercialization from the University of Texas at Austin. He was also the valedictorian of his fire academy class.

Firefighter John E. Gomez was hired in 2006. Firefighter Gomez holds a Bachelor of Science in Finance and Economics with honors from the University of Texas at Dallas. He honorably served in the United States Marine Corps from 1996 to 2000.

Made in the USA
San Bernardino, CA
04 February 2015